Theodore Hoffmann

Ozone or Electro-Negative Oxygen as a Natural Agency in the Vital Process

Its Application in Therapeutics

Theodore Hoffmann

Ozone or Electro-Negative Oxygen as a Natural Agency in the Vital Process
Its Application in Therapeutics

ISBN/EAN: 9783337866129

Printed in Europe, USA, Canada, Australia, Japan

Cover: Foto ©berggeist007 / pixelio.de

More available books at **www.hansebooks.com**

OZONE

OR

ELECTRO-NEGATIVE OXYGEN

AS A

NATURAL AGENCY

IN THE VITAL PROCESS,

ITS APPLICATION IN THERAPEUTICS

BY

DR. THEO. A. HOFFMANN.

BEARDSTOWN, ILL.

& G. A. RAUTH, Steam Printers, 2 Market St. St. Lou

1873.

OZONE.

When after continued dry and sultry weather "in which a sensation of heaviness and debility is creeping upon man, when even the plants hang their leaves, buds and blossoms as if mourning towards the ground", a cool rain and a thundershower sets in, purifying the air nature revives, man breathes more freely, the plants lift their leaves, buds and blossoms towards the bright sunlight, and everything is new life again, such is the effect of "Ozone or Ozonized Air" produced by the electric change and the current in the athmosphere, for which man, animal and plants have languished as a part of their nourishment.

Professor Meissner in Goettingen says : Of all the great leading inventions in chemistry the most important is that of Professor Schoenbein, showing that Oxygen exists in three different forms : 1) as indifferent or chemical-inactive Oxygen, 2) as Antozone or active electro-positive Oxygen, 3) as Ozone or active electro-negative Oxygen. Our atmosphere is composed of 79 Nitrogen and 21 Oxygen with a slight fraction of carbonic acid gas.

The Oxygen of the atmosphere is itself inactive Oxygen, and the activity of the same depends upon the quantity of Ozone it contains, this quantity depending on the electric current in the atmosphere. The richer the air is with Ozone, the more favorable it is for the vital process; the poorer with Ozone, the less favorable it is for our health.

By the process of respiration—Ozone is partly formed in our lungs as an essential agent of our vital process, and the best nourishments would remain inactive for assimilation, if Ozone, a certain quantity of which the saliva carries to the food, were wanting.

Oxygen or Ozonized Oxygen keeps up the warmth in our body, it is the source of force or action for our respiratory and alimentary process. In high plateaux (mountains) the air is richer with Ozone than in low places or crowded cities.

If the quantity of ozonized air to be taken into our system, is reduced to a certain minimum, disease will be the consequence, such as: difficulty in breathing, spasms of the respiratory muscles, sleeplessness, delirium, dilatation of the pupils, asthma, suffocation, paralysis of the heart.—

The effects of the inhalation of ozonized air or ozonized oxygen are mainly: moderate pulse and respiration, increase of appetite, removal of costiveness, acidulated pleasant taste in the mouth, sleep, absorption of exudations.

The inhalation continued for longer periods will produce flushing rosy cheeks and lips, good complexion of the wrinkled skin; hairs which fell out will grow again, pimples and freckles will disappear, the weight of the body will increase; fever, inflammation, spasm, palsy will disappear more or less, or subside entirely. In short, the influence of a purified arterial blood will show its blessings through the whole system of the body, in so far as the sickly affections of the organs can be removed or restored by the influence of a normal blood.

It would be well to remark that an excess either in too often or too long continued inhalation of strong ozonized oxygen will produce nervous irritability, perspiration, insatiable appetite, etc. Therefore the use and the time of inhalation should be under advice of a physician.—Still ozonized air can be introduced into the system without injury, the blood globules absorbing the same. Ozone or ozonized air can not be compared with, or ruled in the same classification as Kreosot or Carbolic acid, which have been tried as purifyers of the circu'ating blood in vain. The cause of the unsuccessful results with those remedies, is that they are in themselves, and in their products of decomposition entirely heterogeneous to the assimilation.

Those suffering with emphysema of the lungs, chronic bronchial catarrh, diseases of the heart, have generally a pale complexion from the venous condition of the blood, and will be very much benefited by the inhalation of Ozone or ozonized air, and life can be considerably prolonged by regular application. In the like manner we can explain the good and successful results, obtained from the inhalation of ozonized oxygen or ozonized air in the treatment of spinal affections, (tabes dorsalis), articular rheumatism, glaucoma, in atrophia of the optic nerv, furunculosis, gout. There might be objections raised by some physicians, and would-be physicians, to the use of Ozone or ozonized air, and when they are founded on real facts they should be acknowledged, still the controversy must not remind us "as Professor Meissner in Goettingen says" of the words:

"What you don't weigh, that has no weight for you;
What you don't coin, you think it is not true."

Eminent physiologists have declared, that the application of gases as effective curatives will have a great future, and in the Ozone or ozonized air we find the great agent, which nature has provided for, (if ever so little) in our atmosphere to be received in our lungs, and if circumstances demand a larger supply for the weary and diseased, it is in our power to satisfy the demand. The diminishing of the contageous influence in epidemics, as cholera, yellow fever, typhus etc., after a heavy thundershower, where Ozone is formed in the atmosphere, speaks for the good effects of the application of Ozone in those diseases.

We observe a curative influence in the neighborhood of Salineworks in scrofula, while travelling on the sea by those suffering with tuberculosis, in staying on high plateaus by persons having intermittent fever, when stiring much in free air by those suffering with chlororis, sickheadache and hoopingcough. This influence is caused by inhaling an air richer on Ozone, than that in dwellings, and the application of Ozone either as inhalation, or drinking of water saturated with Ozone, has given very successful results even in cases where strong doses of quinine could not accomplish the cure.

The light of the sun and electricity are the great agents for sustaining life in the animal as well as in the vegetable kingdom, even the minerals undergo transformation by its influence. Plants and animals depend on oxygen for their existence. Under the influence of sunlight the plants absorb carbonic acid gas from the atmosphere, and exhale oxygen in day-time, so beneficial to men for respiration, during night the plant absorbs oxygen and emits carbonic acid gas.

According to Prof. Traube & Pflüger, the excess of carbonic acid gas is fatal, causing symptoms of suffocation; exactly the same takes place by a deficiency of oxygen.

Carbonic acid chemically combined with ammonia, when accumulated in the blood, will kill with the symptoms similar to those produced when death is caused by freezing.

Of great significance are the cryptogamic spores or microscopic fungies which are floating in the air we inhale ; they play an active part in our vital process. A great part of it may be essential for sustaining the same ; an other part may be inactive or little molesting, another part of it is dangerous to our health by disorganizing our blood, acting as a germ in diseases (generally termed malaria or miasma). Those septical organism : mould, fungie, bastaria, vibriona, zoogloeema etc., or as they may be termed by classification—will accumulate when the conditions for their development in our system are favorable, as for instance, badly ventilated sickrooms, decay and putrifaction of vegetable and animal matter, stagnant water and air, want of sunlight, marshy low places, densely crowded dwellings with accumulation of filth.

Professor Danzer obtained by washing 2495 litres (quarts) of air of the city of Manchester (England) with distilled water a liquid, which contained $37\frac{1}{2}$ millions of cryptogamic spores and fungies. Those 2495 litters air is the average quantity which a middlesized man will inhale in ten hours.—Besides this the air of Manchester contains in 100.000 parts about 25 parts sulphuric acid.

The air is generally deprived of Ozone in all places which are the hotbeds of those miasmatic organisms, and the bad influence of the same may be destroyed by the inhalation of ozonized air. The wise arrangement of nature to prevent a great part of those microscopic bodies or fungies from entering into the circulation, consists in the function of the epithelia and the vibratory and ciliary motion of the same, to take hold or make them adhere to the mucous membrane of the respiratory organs, and throw them out by sneezing or coughing.

Professor W. Kuhne (Schulze's Archiv for microscopic anatomie 11, pages 372—378) has stated that the motion of the cilia will be checked when the air is deprived of oxygen, while their motion is directly revived as soon as oxygen or ozonized air is admitted. That even larger or visible bodies can be carried into our lungs by respiration has been proven by several physiologists.--Professor Traube found fine particles of charcoal in the lungs of persons handling this article extensively.

Tigri, Coze, Houson and Lueders have proven by direct examination, that cryptogermic fungies are carried into the circulation of the blood. The good and fresh complexion of the inhabitants of sunny elevations. an1 the sallow faces of the dwellers in low districts and cellars give a clear state-ment of the influence of oxigen or ozonized air on our system. The quality and the quantity of those septic or miasmatic organisms brought in circulation with the blood, will develop different diseases in our body, as: fevers of various nature, enlargement of the spleen, depression of the nervous system, affections of the respiratory organs, ros-eola (rosetrash), initial bleeding. diphtheritis, cholera, puerperal fever. yellow fever, dysenteria, catarrhus, cerebro spinal meningitis, paralysis, exanthemic diseases etc.

The antiseptic qualities of quinine are well known and acknowledged, and it is an erroneous view to ascribe the origin of dropsy, softening of the brain, enlargement of the spleen and liver, to the use of quinine, which the practical observations by the writer (author) for the last 4 decennaries can confirm. Those diseases appear frequently, when the character of the fever is suppressed by small or insufficient doses of quinine, yet the influence of the malaria or the septic germs in the blood not yet perfectly removed, and will be the cause of congestion in one or the other part of our body, which generally yields to the application of energetic doses of quinine in proper shape.

We have in Quinine and Ozone two powerful antagonistic remedies to check the developement of those miasmatic cryptogermic spores, fungics or septical germs, to destroy the same or make them harmless. While those germs exist and develop at the expense of the oxygen in our blood, an addition of ozonized air into the circulation of the blood will restore the same to its normal state. Professor Binz considers quinine as a strong ozonizing agent.

A current of ozonized air conducted into water, will make the infusorial animals (monada) at first motionless and afterwards destroy them entirely. A current of ozonized air into a foetid liquid blood will restore the original smell.

Concentrated Ozone is not well applicable for inhalation, being too energic, in the same manner as alcohol of 95 degrees is not fit to be applied as medicine without dilution. Ozonized air should not be stronger, than that it may be inhaled without producing cough; a grown man might inhale from 25 to 100 quarts per day without danger.

100 litres (quarts) atmospheric air contains about 0,01 to 0,02 milligramme Ozone. (Prof. Ludwig). According to Prof. Genth, Ozonized air will give the best relief to persons suffering with gout. Prof. Ludwig Boehme considers cholera a desquamative acute catarrh of the intestines accompanied with microscopic fungics. An application of ozonized air by the mouth, and ozonized water concentrated as injection, might give the practitioner a more effective remedy on hand than all others known.

In poisoning with gases, such as sulphuretted hydrogen, carbonated oxygen gas and others similar. The application of Ozone should be preferred before any other remedy.

Pristly & Scheele discovered the oxygen anno 1774. Five cubicfeet atmospheric air contain one cubicfoot oxygen and four cubicfeet Nitrogen. Oxygen has a greater specific gravity than atmospheric air.

The density of Ozone is 1.50 to oxygen. Oxygen will remain in a gaseous form by a pressure of 58 atmospheres, and a temperature of 139 degrees Fahrenheit below Zero. Farraday. 100 volumes of water will absorb 3 to 4 vol. oxygen by the height of the barometer of 760 cent. metr. Ozonized air or oxygen changes the color of the dark venous blood into high red. The same change takes place when animals inhale oxygen gas; and one vol. of oxygen gas will sustain life four times longer than the same vol. of atmospheric air.

According to Schoenbein, manganic acid and chromic acid contain electro-negative oxygen or ozone, therefore called by him ozonides. Ozone mixed with athmospheric air will produce no inflammation of the lungs. According to the same author and others, Albumen milk, the fibrine and globules of the blood will absorb Ozone freely. Also powdered carbon (charcoal).

In the city of Paris (France) are 19 stations, where Ozone contained in the atmosphere is measured.

According to Alex. Schmidt. The normal blood actually contains Ozone.

Ozonized air has been inhaled for twelve weeks with comfort and without effects of coughing or signs of inflammation. Ozonized oxygen will effect Indian Rubber, and it will soon loose its good effects, when kept in indian rubber bags. The best way to use Ozone for inhalation is in its nascent state.

According to Prof. Kuehne. The blood globules are carriers of Ozone. and act on tests, such as iodide of potassium, tinct. of guajac.

Bile absorbs Ozone energetically.

The importance of the inhalation of ozonized air in jaundice, gout, oxalic acid diathesis, is mentioned by Goromp and Besanog.

According to R. Richter. arrow poison in watery solution is decomposed by a current of ozonized air.

According to Blanche. Taddei. Schoenbein. Plants contain Ozone.

According to Garrod. the inhalation of 28 litters of ozonized air twice a day, produced relief in acute gout. Eckhardt considers the inhalation of ozonized air in gont as sufficient without any other remedy. C. Paul, Husemann, Herman and others recommend the inhalation of Ozone in cases of poisoning with opium, sulphuretted hydrogen, carbonic oxyd gas. etc.

Ferd. von Arnim, inhaled for two days 150 quarts of ozonized air each day, for the two successive days 180 quarts each time, and the fifth day 250 quarts with good effects.

The origin of Ozone or ozonized air in our atmosphere is stated in the foregoing. The quantitative variation of it depends on the meteorological influence, and in certain circumstances this quantity of Ozone may be reduced to Zero. Ozone or ozonized oxygen is formed in the electric current of a galvanic or electric battery in diverse operations, representing the electro-negative oxygen.

It is also formed in various chemical actions. The difficulties of the diverse ways to produce Ozone cheaply and conveniently have checked its practical application for

general use more or less. By my lately patented process
I have succeeded to produce Ozone or ozonized air in a
cheap and practical manner, to enable physicians or med-
ical men to apply the same, whenever circumstances demand it in its nascent state, mild or strong, or in local application as Spray by attaching an atomizer.

The foregoing statements of expert practitioners and
eminent Physiologists, will give encouragement to the application of this great and valuable remedy, which nature
has provided for the vital process under the laws existing
from the beginning of the world, yet unknown to men, before we were advanced in science far enough to decypher
gradually the concealed and secret rules existing in the
great immensurable universe, and so we advance step by
step in the developement of science for the benefit of man-
kind, that will give us a right to claim that man is the
most superior and most important being dwelling on this
insignificant but eventful and progressive world.

APPENDIX.

The peculiar quality. that albumen or albuminous bodies absorb Ozone freely, has given impulse to the technical application of Ozone to fresh fermented liquids, after the same have got through what is technically called: "the first fermentation and separated from the yeast." The wines in this state even if clear in appearance, have still albuminous bodies in solution, which in the warm season will produce a second fermentation. Those bodies will be oxydized by Ozone, and more easily separated, and thus the wine ripens in a shorter period, acquiring at the same time a finer bouquet (flavor). The good results of this practical application. and the easy manner in which it can be performed, will give this method the preference to that of "Pasteurs", which consists in heating the wines to a certain degree in order to destroy the effects of those fermenting bodies, by which however a part of the fine flavor will by destroyed also. Alcoholic liquids or distilled liquors treated with Ozone will improve in a short time in the same manner as by age if stored in warehouses for several years in barrels with open bungholes so as to absorb the small quantity of Ozone in the air, losing in this way considerable, by evaporation, in quantity, also the interest on the invested capital.